Beans

by Gail Saunders-Smith

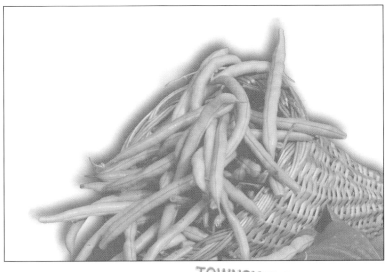

Pebble Books

an imprint of Capstone Press

Pebble Books

Pebble Books are published by Capstone Press
818 North Willow Street, Mankato, Minnesota 56001
http://www.capstone-press.com
Copyright © 1998 by Capstone Press
All Rights Reserved • Printed in the United States of America

Library of Congress Cataloging-in-Publication Data
Saunders-Smith, Gail.
  Beans/by Gail Saunders-Smith.
  p.cm.
  Includes bibliographical references (p. 23) and index.
  Summary: In simple text and photographs describes planting, growing, picking and eating green beans.
  ISBN 1-56065-487-2
  1. Beans--Juvenile literature. 2. Beans--Life cycles--Juvenile literature. [1. Beans.] I. Title.

SB327.S28 1997
635'.652--DC21
              97-23585
                CIP
                AC

Editorial Credits

Lois Wallentine, editor; Timothy Halldin and James Franklin, design; Michelle L. Norstad, photo research

Photo Credits

Michelle Coughlan, 4
Winston Fraser, 10
Dwight Kuhn, cover, 3 (left), 6, 8
Unicorn Stock/Joel Dexter, 12; Eric Berndt, 14; Ted
  Rose, 1, 3 (right), 16; Tom Edwards, 18; Karen
  Holsinger Mullen, 20

2

Table of Contents

3

beans in a packet

beans in a seed

beans in a hole

beans in a garden

beans in a blossom

beans in a pod

beans in a basket

beans in a bowl

beans and me

Words to Know

bean—a vegetable with large pods

blossom—the flower on a plant that becomes a fruit or vegetable

garden—a place outside where plants grow

packet—a small envelope for seeds

pod—the shell around the seed of the bean

seed—the part of a plant from which a new plant can grow

Read More

Gibbons, Gail. *From Seed to Plant.* New York: Holiday House, 1991.

Jennings, Terry. *Seeds.* New York: Gloucester Press, 1988.

Wexler, Jerome. *Flowers, Fruits, Seeds.* New York: Prentice-Hall Books for Young Readers, 1987.

Internet Sites

Chucks Produce Talk
http://www.comevisit.com/chuckali/produce.htm

Dole 5 a Day
http://www.dole5aday.com

Welcome to the Internet Food Channel
http://www.foodchannel.com

Note to Parents and Teachers

This book describes and illustrates planting, picking, and eating green beans. The text and photographs detail the growth cycle and sequence of events that occur from planting beans through eating beans. The photographs clearly illustrate the text and support the child in making meaning from the words. Children may need assistance in using the Table of Contents, Words to Know, Read More, Internet Sites, and Index/Word List sections of the book.

Index/Word List

Word Count: 35